SCIENTIFIC AMERICAN | EDUCATIONAL PUBLISHING

SCIENTIFIC AMERICAN INVESTIGATES REPTILES

GECKOS

BY JAGGER YOUSSEF

Published in 2026 by The Rosen Publishing Group
in association with Scientific American Educational Publishing
2544 Clinton Street, Buffalo NY 14224

Cataloging-in-Publication Data
Names: Youssef, Jagger.
Title: Geckos / Jagger Youssef.
Description: Buffalo, New York : Scientific American Educational Publishing, an imprint of Rosen Publishing, 2026. | Series: Scientific American investigates reptiles | Includes glossary and index.
Identifiers: ISBN 9781725352858 (pbk.) | ISBN 9781725352865 (library bound) | ISBN 9781725352872 (ebook)
Subjects: LCSH: Geckos–Juvenile literature.
Classification: LCC QL666.L245 Y68 2026 | DDC 597.95'2–dc23

Portions of this work were originally authored by Kathleen Connors and published as *Geckos*. All new material in this edition is authored by Jagger Youssef.

Designer: Andrea Davison-Bartolotta
Editor: Therese Shea

Photo credits: Cover, p. 1 Miroslav Halama/Shutterstock.com; p. 5 Tony Moran/Shutterstock.com; p. 7 (top) Lauren Suryanata/Shutterstock.com; p. 7 (bottom) Kurit afshen/Shutterstock.com; p. 9 Sebastian Janicki/Shutterstock.com; p. 11 Mr.B-king/Shutterstock.com; p. 12 Eric Isselee/Shutterstock.com; p. 13 anne-tipodees/Shutterstock.com; p. 15 Spok83/Shutterstock.com; p. 17 Landshark1/Shutterstock.com; p. 19 (top) Fotografie-Kuhlmann/Shutterstock.com; p. 19 (bottom) Matt Jeppson/Shutterstock.com; p. 21 (bottom) reptiles4all/Shutterstock.com; p. 21 (chart) alinabel/Shutterstock.com; p. 21 (gecko in chart) IGORdeyka/Shutterstock.com.

Some of the images in this book illustrate individuals who are models. The depictions do not imply actual situations or events.

Printed in the United States of America

CPSIA compliance information: Batch #CSSA26. For Further Information contact Rosen Publishing at 1-800-237-9932.

Find us on 

CONTENTS

Words in the glossary appear in **bold** type the first time they are used in the text.

CLIMBING CREATURES

Geckos are lizards known for their amazing climbing skills. They can climb on walls and even ceilings! Like all reptiles, geckos are **cold-blooded**. They have a backbone, are covered with dry, scaly skin, and usually lay eggs.

There are more than 1,000 species of geckos. Most are less than 6 inches (15 cm) long. That includes their tail. They have four short legs and **stout** bodies. Geckos can be several colors, such as gray, brown, green, and whitish.

The largest species of gecko is the New Caledonian gecko, shown here. It can grow to be 14 inches (36 cm) long.

LOOK INTO THEIR EYES

Most gecko species are nocturnal. That means they're most active at night. These geckos have **pupils** that are different from geckos that are diurnal, or most active during the day.

Nocturnal geckos have **vertical** pupils. This helps them hunt in low light. Diurnal geckos have round pupils. All geckos have good eyesight.

Most geckos have no eyelids! Instead, they have a clear scale over their eye. When geckos shed, or get rid of, their scales, they may clean their eyeballs with their tongue until a new scale forms!

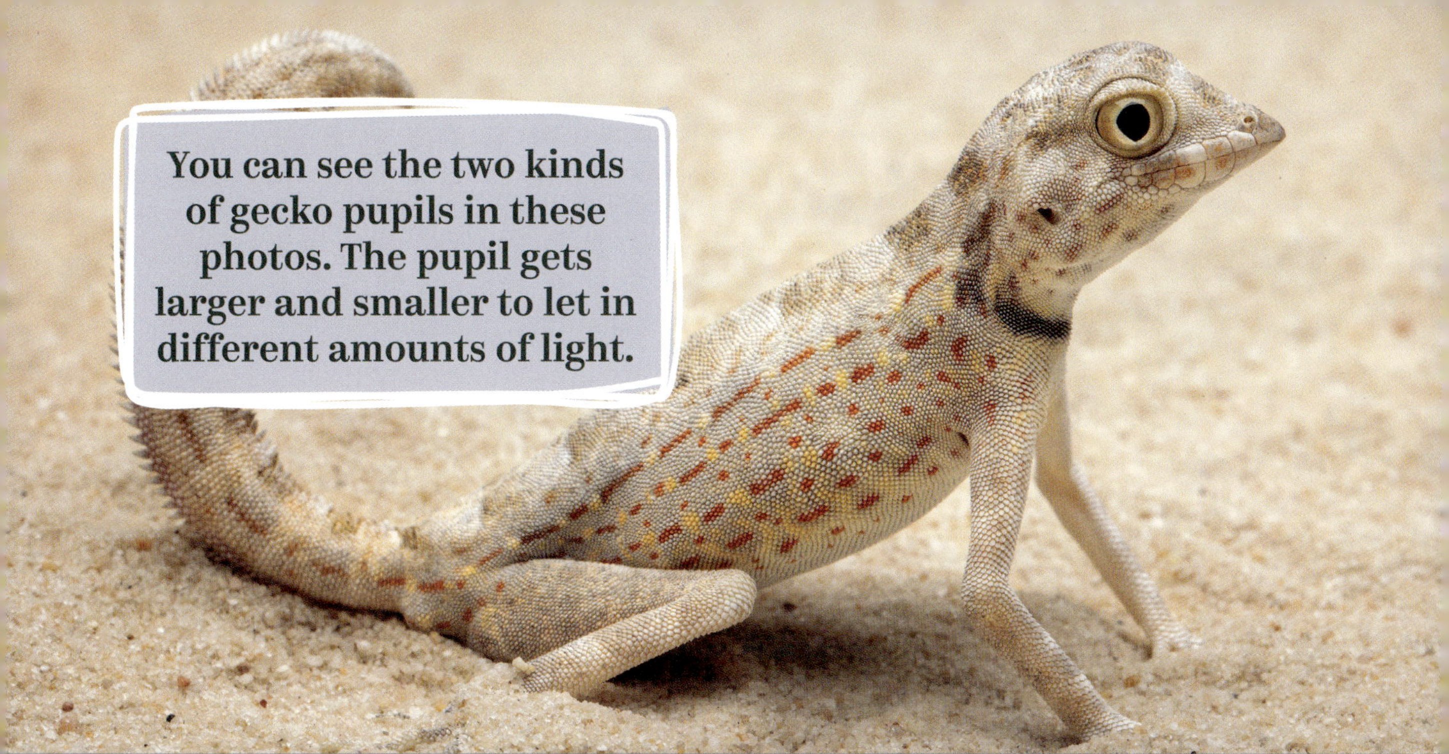

You can see the two kinds of gecko pupils in these photos. The pupil gets larger and smaller to let in different amounts of light.

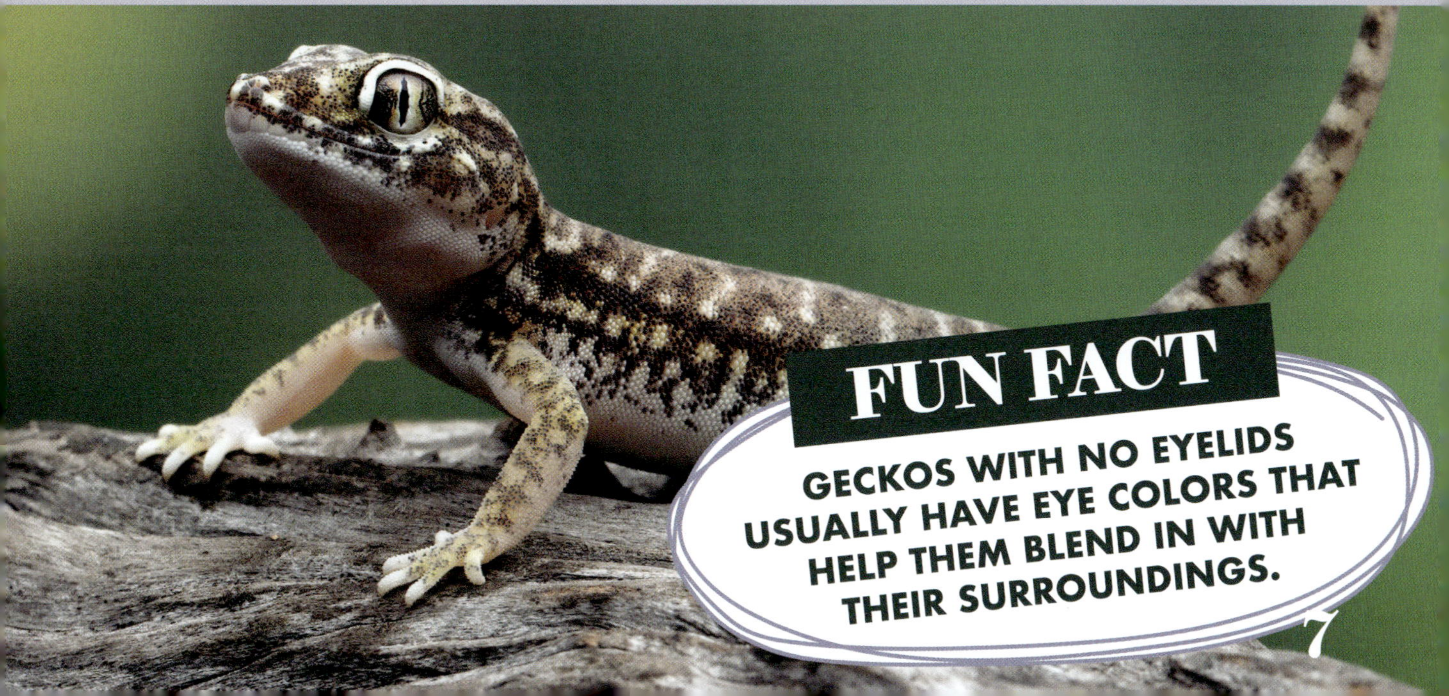

FUN FACT

GECKOS WITH NO EYELIDS USUALLY HAVE EYE COLORS THAT HELP THEM BLEND IN WITH THEIR SURROUNDINGS.

SENSING THEIR SURROUNDINGS

A gecko's ears are tiny tunnels on both sides of their head. Its ears help it sense the direction of noises.

Geckos use both their nose and tongue to smell. A gecko sticks its tongue out to sense smells. The tongue carries particles of odor to the Jacobson's organ, which is a small group of cells on the roof of a gecko's mouth. The Jacobson's organ can sense prey as well as **chemical** messages. For example, geckos release, or let go of, chemicals as a message if they're looking for a **mate**.

Geckos have tiny parts in their ears called saccules. Saccules help them feel **vibrations** in their surroundings.

FANTASTIC FEET

Fast-moving geckos can be hard to catch—especially if they're on the ceiling! How do these reptiles do this?

Most geckos have toes covered with plates that have many tiny hairs. The hairs act as little hooks to help the creatures climb. In addition, scientists think geckos use forces in nature to "stick to" smooth surfaces. These include **friction** and **van der Waals force**. Some geckos have claws to help them get around too. Not all geckos have the same amazing climbing skills, however.

Scientists have discovered that gecko toes have sticky lipids, or fats, on them.

FUN FACT
AT LEAST ONE SPECIES OF GECKO CAN RUN ACROSS WATER WITHOUT SINKING!

LISTENING TO GECKOS

Geckos are the loudest lizards. They click, chirp, and squeak to communicate with each other. Some of the sounds they make can't be heard by people.

The sounds a gecko makes depends on its species. For example, a New Caledonian gecko growls. The turnip-tailed gecko makes clicks like an insect. The tokay gecko might hiss or croak at an attacking predator. The thick-tailed gecko defends its territory with a sharp, barking noise. Barking attracts mates for some species too.

The thick-tailed gecko, shown here, is also called a barking gecko for the sounds it makes. It's only found in Australia.

FINDING MATES

Geckos usually like to live alone, in their own territory. However, they will try to find other geckos when it's time to mate. The male gecko will sound his mating call to let a female know it's safe to come closer. During the rest of the year, geckos guard their territory against any animal—including other geckos—that might eat up their food.

After mating, female geckos commonly lay two white eggs at a time. Some species lay eggs once a year. Others **reproduce** more often than that.

A baby leopard gecko comes out of an egg. It's called a hatchling.

FUN FACT

A FEW GECKO SPECIES IN NEW ZEALAND GIVE BIRTH TO LIVE YOUNG INSTEAD OF LAYING EGGS.

GROWING UP GECKO

Gecko mothers often lay their eggs under tree bark or leaves. Then, they usually leave the eggs alone. The size of hatchlings depends on the species. They often look like tiny copies of adult geckos. These little lizards don't have to look for food right after hatching. They shed their first layer of skin and eat it!

A gecko's lifespan varies greatly between species. Geckos usually live less than 10 years in the wild. But some in zoos and homes can live as long as 20 years.

Eating shed skin provides **nutrients** geckos need to grow.

GECKOS AT HOME

Geckos live on every continent except Antarctica! They're usually found in warmer, humid, or wet, places. They live in rainforests, deserts, mountains, and other kinds of habitats, or natural homes. Many geckos like to live in trees. Geckos that live in deserts may dig **burrows**.

Geckos eat mostly bugs, though they may eat fruits and plants too. Some large species eat other lizards. If geckos live in your house, you've got built-in pest control. Geckos eat moths, cockroaches, and other bugs that people don't want in their house.

The web-footed gecko, which has nearly see-through skin, lives in Africa's hot, dry Namib Desert.

FUN FACT
A GECKO CAN DROP ITS TAIL IF A PREDATOR TRIES TO CATCH IT. THE TAIL KEEPS MOVING AS THE GECKO TRIES TO GET AWAY. MOST SPECIES CAN GROW A NEW TAIL.

CARING FOR GECKOS

Many species of geckos make good pets—if they're cared for well. They need an aquarium with a covering, a place to climb, and a source of heat. They usually don't like to be handled.

Some species of geckos are endangered, which means in danger of dying out. The Union Island gecko almost died out because they were taken from the wild to sell as pets. Other species are endangered because people take over their wild homes. Keep learning about geckos. They're important parts of their habitats!

A Gecko's Food Chain

Gecko predators include snakes, birds, and spiders.

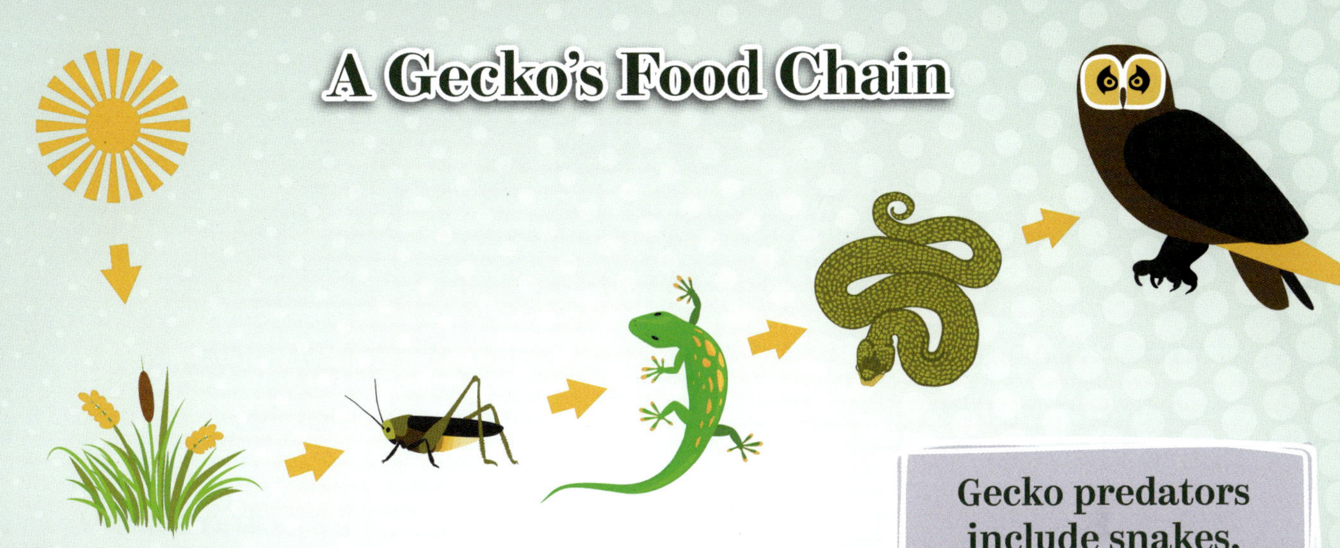

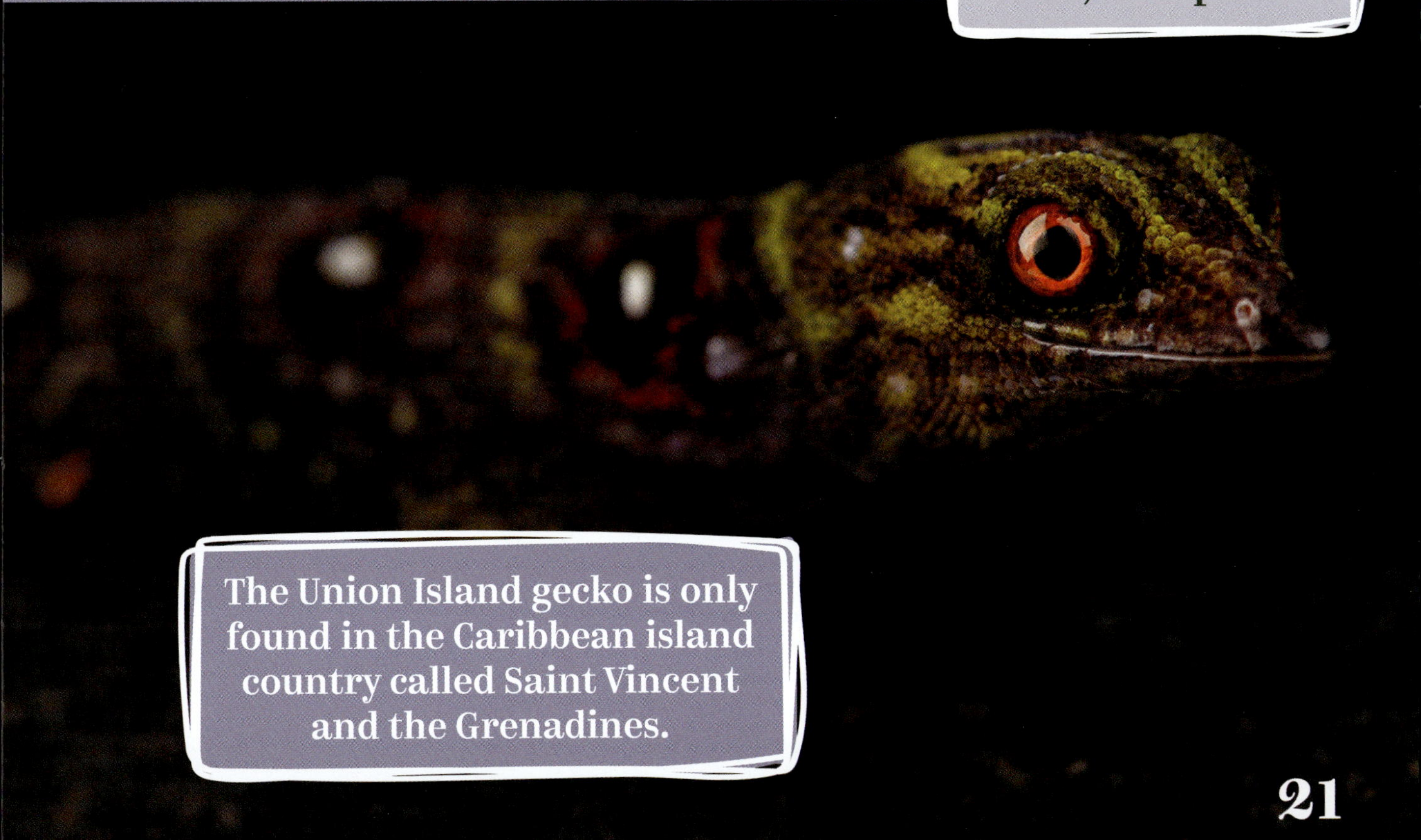

The Union Island gecko is only found in the Caribbean island country called Saint Vincent and the Grenadines.

GLOSSARY

burrow: A hole made by an animal in which it lives or hides.

chemical: Matter that can be mixed with other matter to cause changes.

cold-blooded: Having a body temperature that's the same as the temperature of the surroundings.

friction: The force that resists motion between two things that are touching.

mate: One of two animals that come together to make babies. Also, to come together to make babies.

nutrient: Something a living thing needs to grow and stay alive.

pupil: The black part in the center of an eye that takes in light.

reproduce: When an animal creates another creature just like itself.

stout: Heavy build.

van der Waals force: A weak force that draws different particles called molecules together.

vertical: Straight up and down.

vibration: Rapid movements back and forth.

FOR MORE INFORMATION

Books

Maloney, Brenna. *Gecko or Komodo Dragon*. New York, NY: Children's Press, 2024.

Riggs, Kate. *Geckos*. Mankato, MN: The Creative Company, 2023.

Websites

Geckos

kids.nationalgeographic.com/animals/reptiles/facts/gecko
Read more about geckos from National Geographic Kids.

Tokay Gecko

nationalzoo.si.edu/animals/tokay-gecko
The National Zoo has lots of fun facts about this gecko species.

INDEX